Practical Action Publishing Ltd
27a Albert Street, Rugby, CV21 2SG, Warwickshire, UK
www.practicalactionpublishing.org

First published 1995\Digitised 2013

ISBN 10: 1 85339 270 7
ISBN 13: 9781853392702
ISBN Library Ebook: 9781780443669
Book DOI: http://dx.doi.org/10.3362/9781780443669

A catalogue record for this book is available from the British Library.

Since 1974, Practical Action Publishing has published and disseminated books and information in support of international development work throughout the world. Practical Action Publishing is a trading name of Practical Action Publishing Ltd (Company Reg. No. 1159018), the wholly owned publishing company of Practical Action. Practical Action Publishing trades only in support of its parent charity objectives and any profits are covenanted back to Practical Action (Charity Reg. No. 247257, Group VAT Registration No. 880 9924 76).

List of contents

Introduction

Soybeans (Glycine max) are an important component of the diet in South East Asia where on average about 77g are consumed per person per day, in various forms which include both fermented and non-fermented products. The protein in soybeans is of high biological value and the amino acid composition complements that of cereals (for example, the lysine content of soybean is 6.9 and of rice is 3.8g/16gN, and the methionine content of soybean is 1.6 and of rice is 3.4g/16gN). Hence their importance as protein sources in Asia, where they provide 10 per cent of the total protein intake and where rice is a staple food.

Chinese beancurd (doufu) became popular more than two thousand years ago during the Western Han Dynasty and was later introduced into other countries such as Korea and Japan who in turn developed their own products. According to ancient traditional Chinese medicine[1], beancurd is known 'to prevent coughs, eliminate phlegm, stop inflammation, be beneficial to the stomach and remove excessive heat in the lungs'.

The aim of this booklet is to review the wide range of beancurd varieties available, the principles used in their manufacture by traditional methods and those methods that have potential for large-scale manufacture. The information gathered was obtained by discussions with local doufu manufacturers and by consulting Chinese literature, including patents.

Figures 1a and 1b show the main beancurd products and their composition. The product names are given in Pinyin (phonetic) form.

Figure 1a. Summary of the main soybean products in China

Type	Name	Details	Uses
a) Fermented	Soy sauce (jiang you)	Salty brown sauce derived from fermentation with Aspergillus, Lactobacilli and yeasts for up to one year.	Used to season dishes during cooking.
	Soy paste (jiang)	Brown paste fermented for one to three months by yeast and lactic acid bacteria.	As above.
	Fermented beancurd (furu)	Cubes of beancurd fermented with Mucor and packed in brine. Flavour resembles soft, strong cheese.	Used during cooking or eaten cold along with a dish as a condiment.
b) Non-fermented	Soya beans (dadou)	Tender, unripe beans (two or three per pod).	Cooked for 15 minutes, podded and eaten as a vegetable or deep-fried, spiced and eaten as a snack.
	Beanspouts (douya)	Beans germinated for five to ten days to give long, thick sprouts. Trypsin inhibitor inactivated during sprouting.	Stir fried as a vegetable; pickled.
	Soya bean milk (doujiang)	Liquid produced by mashing soaked beans, cooking and filtering. Nutritive quality is 60-90% of cows milk. May be spray or drum dried.	Served as a beverage, hot or cold. Flavoured eg chocolate, banana, coconut.
	Bean powder (doufen)	Ground, roasted beans.	Used in baked foods.
	Beancurd (shui doufu)	Shiny, smooth white curd from coagulated soya bean milk. May be further treated by smoking, drying, deep-frying etc.	Important protein source. Usually stir fried with seasoning. May be steamed or braised.
	Dried curd bamboo (fuzhu)	The yellow skin which appears on the surface of boiling soybean milk before the coagulant is added. Rolled and dried as sticks.	Rehydrated in warm water and eaten cold, fried or boiled with seasonings.
	Dried curd skin (fupi)	As above but dried as folded discs.	Rehydrated as above and used to wrap foods especially for banquet dishes.
	Wrapped beancurd (bao doufu)	Made by wrapping individual squares of curd in cloth before pressing so they have rounded edges. Sold in tubs of water.	Delicate flavour so eaten steamed.

Figure 1b. Proximate composition of soybean products (per 100g)

	Water	Protein	Fat	Carbohydrate	Fibre
Fresh soya bean	68	13	6	11	2
Soy sauce	72	7	0.5	2	0
Soy paste	50	14	5	16	2
Fermented beancurd	60	17	14	0.1	0.1
Beansprouts	82	8	2	8	1
Soya milk	93	3.4	1.5	1	0.4
Bean powder	5	38	19	32	3
Beancurd	87	7	3.5	3	0.1
Dried curd bamboo/sheets	9	52	24	12	0.1
Soya bean flour	2.3	39	21	31	6.5
Defatted soya bean flour	2.3	46	5	38.4	8.1

Some figures are based on (2).

Manufacture

Chinese beancurds are still being produced by traditional methods at more than one hundred and fifty thousand small family run factories, which are often one room of the family's residence. The products are sold at the premises or transported by bicycle and sold at the local free-markets.

The various types of beancurd are made in two stages: first the soya milk is prepared and secondly the soya milk is coagulated to form curds which are then pressed. Further treatment of these curds gives rise to the many different types of product. The processing of soybeans is necessary to inactivate the antinutritional substances, to avoid the unacceptable beany flavour and to increase their digestibility. Soybeans are soaked before being mashed in water to a mixture which is then filtered to remove the solids. The liquid part is cooked to detoxify it and a coagulating agent is finally added to separate the curds. The curds are sold fresh or processed to alter the texture by pressing, frying, drying and fermenting.

Preparing the press.

The standard white beancurd is made according to the flow diagram shown in Figure 2.

Figure 2. The manufacture of standard beancurd

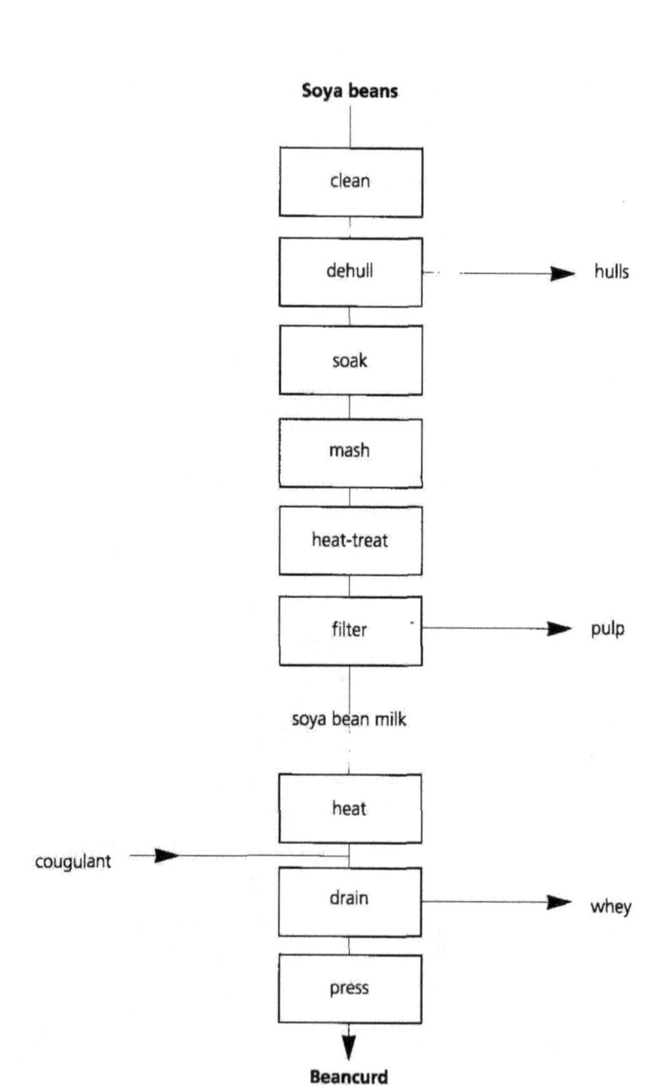

Filtration of the mash.

a. **Cleaning and Soaking.** The beans are sieved to remove debris such as stones and then washed and dehulled. The beans are soaked in water at a ratio of beans to water of one to two. The length of the soaking stage depends on the season (see Figure 3). The soaking stage removes most of the bitter substances present in the beans.

b. **Mashing.** Sufficient water is poured onto the soaked beans to cover them and the mixture is ground to a mash.

c. **Heat Treatment.** Boiling water is added to the mash in the ratio of water to mash of 1.5/2:1 and the container is left covered for 15 minutes. An alternative is to cook the mash under pressure.

d. **Filtration.** The liquid obtained after the heat treatment is poured into a large cotton bag which is pressed by hand. The pulp, which remains inside the bag, is mainly fibre and crude protein and is usually used for animal feed but also occasionally for humans. The liquid is the useful soybean milk.

e. **Heating and Coagulation.** The soybean milk is boiled. A small amount of coagulant is placed into a wooden or metal container and the hot milk is added. Finally the remainder of the coagulant is added and the mixture left for 15-20 minutes to coagulate.

Figure 3. Table to show soaking times during different seasons

Season	Room temperature (°C)	Water temperature (°C)	Soak time (hours)
Winter	0	5	20-22
Spring	10	10	12-14
Autumn	24	20	10-11
Summer	30-40	25-30	6-7

f. **Draining and Pressing.** Unbleached cheesecloth is used to line a wooden container (typically pine or cedar) into which the coagulated milk is carefully poured. The remainder of the cloth is folded on top and a wooden lid is fitted. Different weights give varying amounts of pressure and therefore produce curds of different moisture contents and textures. For soft doufu the pressures used are: 2-4g/cm^2 for five minutes then increased to 5-10g/cm^2 for 10-15 minutes or until the whey stops draining out.

Vessels to collect and heat filtrate.

For a firmer curd the pressure required is 20-100g/cm^2 for 20-30 minutes. The final curds may be cooled in water at 5°C for 60-90 minutes before being packaged and refrigerated or sold directly.

The advantages of this method are that once it has been mastered it is relatively simple to carry out, it can be used and afforded by the average family and the equipment is simple. The main disadvantage, apart from the time and labour required, is that the yield is quite low (eg 1kg of dry soybeans gives 9kg of fresh beancurd) and the amount of by-product, which yields very little income, is high. However, research is being carried out to increase the use of the by-product in human foodstuffs. The composition of the soybean curd is shown in Figure 4.

Figure 4. Composition of soybean curd (per 100g)

Protein	5-8g
Fat	3-4g
Carbohydrate	2-4g
Water	84-90g
Fibre	0.1g
Energy	0.32MJ
Calcium (Ca)	1.0mg
Iron (Fe)	1.8mg
Phosphorous (P)	0.95mg
Vitamin B1	0.05mg
Vitamin B2	0.04mg
Nicotinic acid	0.5mg

Packaging

The curd is soft and delicate and must be handled carefully to avoid disintegration. It is often sold outside the place of manufacture on the wooden lid which serves as a tray. The construction of the box in which it is pressed is designed to leave ridges on the surface (Figure 5). These serve as guidelines for the salesperson who cuts the slab into pieces and puts them directly into the customer's container or into thin polythene bags. The beancurd is sold per piece not by weight.

Figure 5. Removing the curd

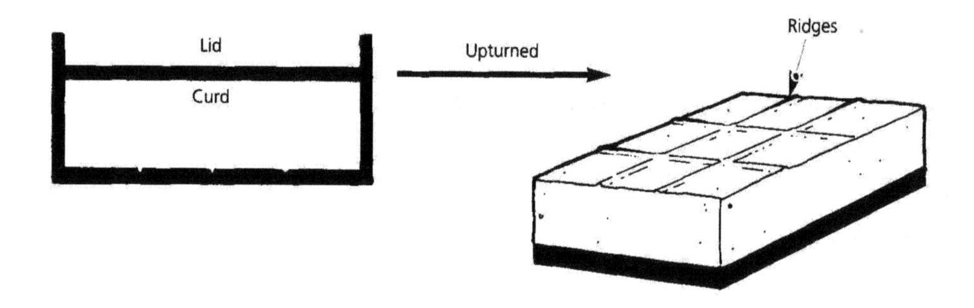

The curd should be stored under water to prevent it drying out and to prevent the formation of a brownish discolouration on the surface. To extend the shelf life it may be packaged in different ways:

- In plastic bags for 7-14 days if refrigerated.
- Heat sealed in water filled plastic trays, possibly with some flavour losses.
- Vacuum packed for 3-5 weeks.
- Sold in bulk at 4°C. The water must be changed daily.
- Packed in jars, heated for one hour in boiling water and given an airtight seal, it has a shelf life of six months (3).

Equipment

The basic equipment varies slightly with different regions and manufacturers but the equipment traditionally used in mainland China is explained below.

Figure 6. Sieve

Sieves are used to remove foreign matter and wash the beans.

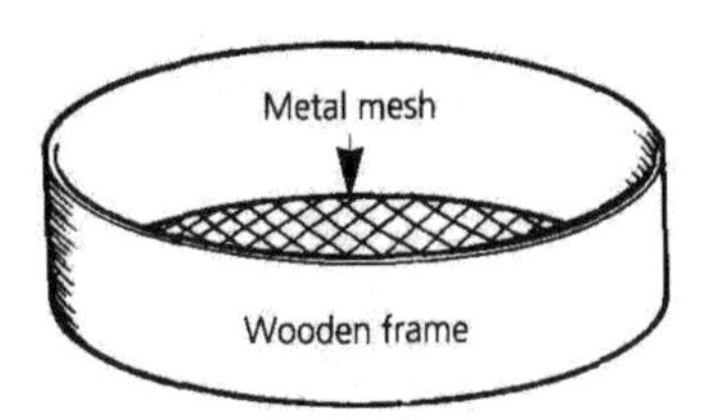

Figure 7. Stone mills

Traditionally, hand-operated granite millstones are used to grind the beans, although some may now be fitted with electric motors to speed up the operation. They give a fine puree, high yields and are hygienic. The fabrication of the stones is important for correct grinding and about 500g of beans are put on the millstone surface.

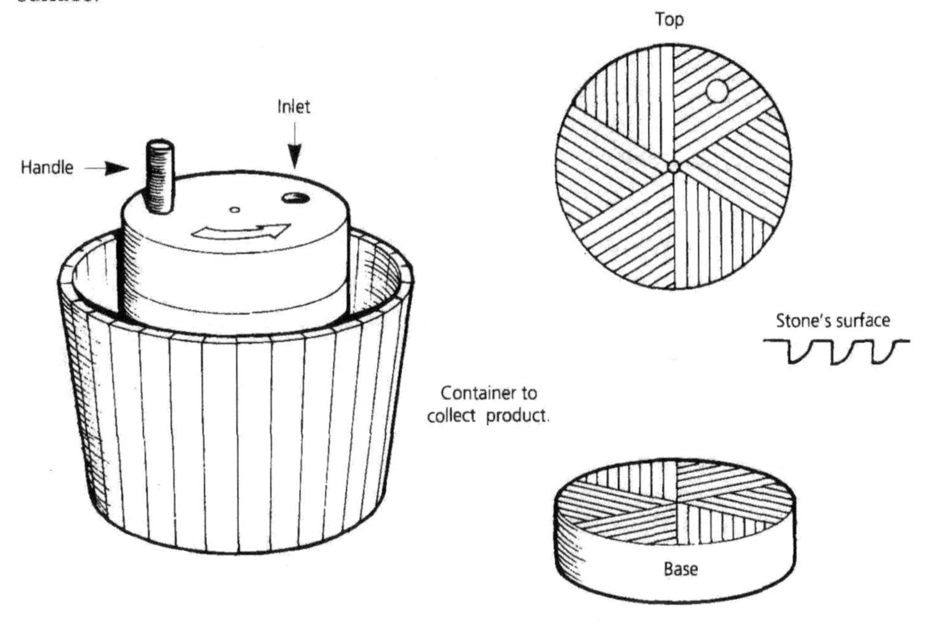

Figure 8. Soaking tank

Large stone urns are filled with water using a hose-pipe connected to a tap.

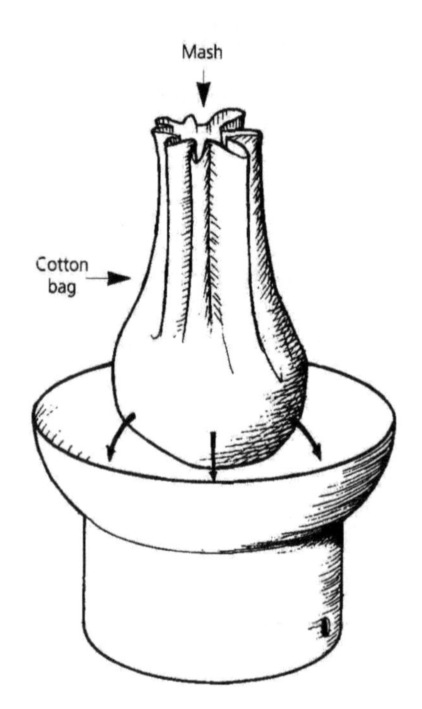

Figure 9. Filtration

The mash is collected in a container, boiling water is added and is poured into a cotton bag and the whole is suspended from the ceiling of the processing room. The bag is squeezed by hand and the liquid drains into a collecting vessel, such as a very large wok, which can then be heated on a fire.

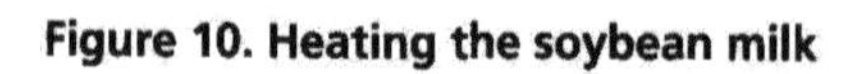

Figure 10. Heating the soybean milk

The wok is covered and heated.

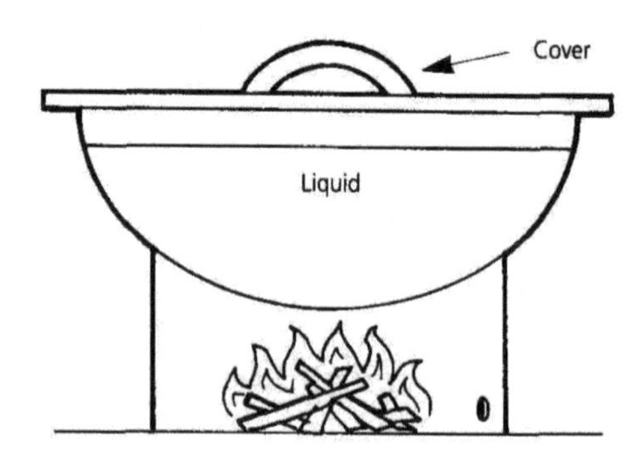

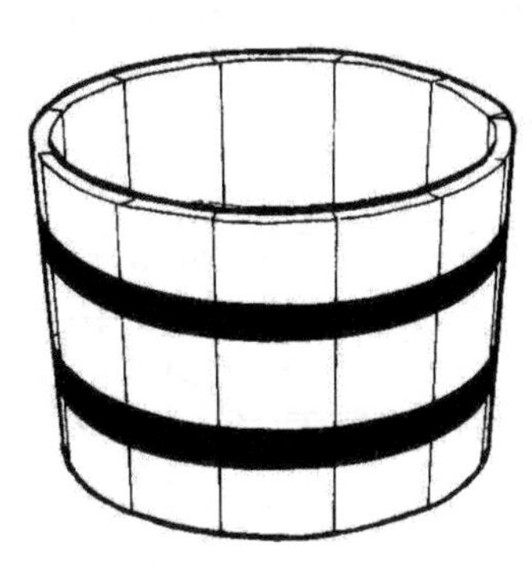

Figure 11a Coagulating equipment:
and b. Pestle and mortar and wooden container

Gypsum is heated and ground by hand with a pestle and mortar to form calcium sulphate. A small amount is placed in the container before the soya milk is added followed after filling by the remainder of the coagulant. The liquid is covered and left to stand.

Curd pressed by wrapping in cloth.

Figure 12a. and b. Construction and use of the wooden box for pressing

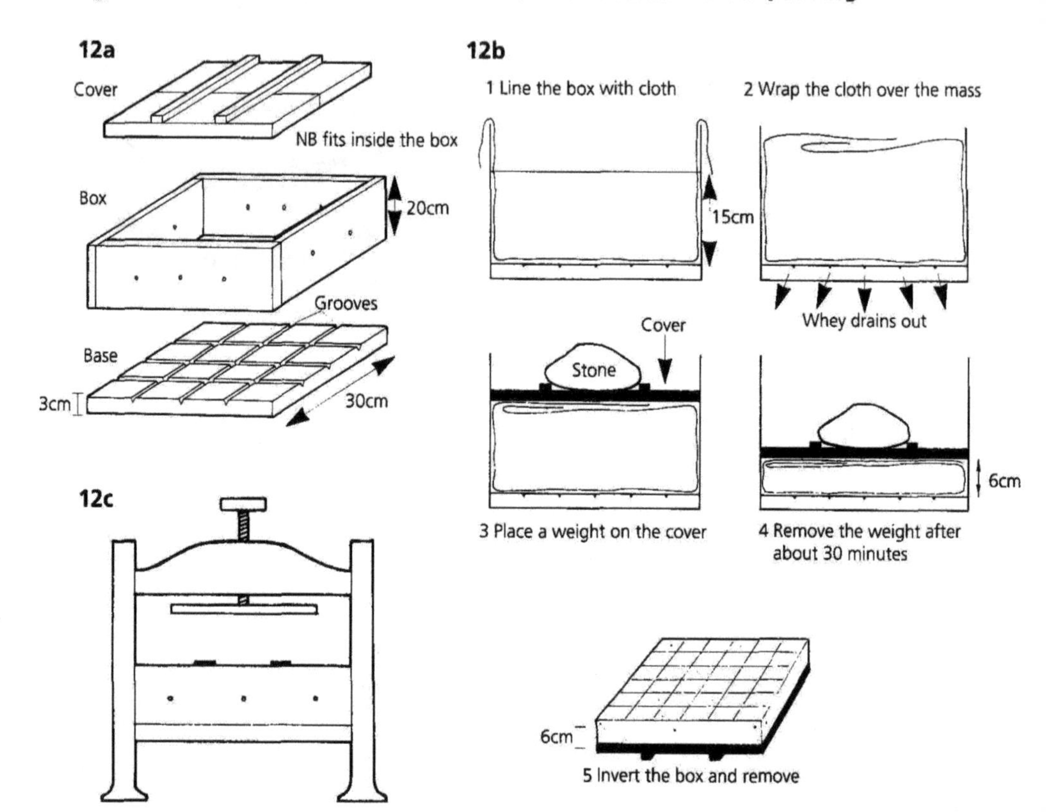

Figure 12c. Hand operated press

A specially designed square wooden box with a separate lid and base is used. The sides and lid of the box have holes to allow the whey to drain out and 3-5 grooves are cut at right angles into the base, which also allows drainage. The box is lined with a double layer of cheesecloth which is wrapped carefully to cover the mass. Pressure is applied by a hand turned press or by placing weights on the lid so that after about 30 minutes the thickness of the product is reduced to 6cm.

Grid for dividing into small squares.

Pressing the curd.

Coagulating agents

The coagulation step is important and requires care. If the curd has a crumbly texture, it is likely to have been caused by adding too much coagulant too fast and at too high a temperature, causing the curds to form too quickly. Insufficient coagulant will give a low yield.

Different coagulants are used either singly or as a mixture. For example, in southern China, calcium sulphate is more common whereas in the north, magnesium chloride is used to give the preferred firmer curd. The different types of coagulation used can be classified as follows:

a. **Nigari Type.** This is produced from sea water during salt manufacture and contains magnesium sulphate ($MgSO_4.7H_2O$), magnesium chloride ($MgCl_2.6H_2O$) and calcium chloride ($CaCl_2$). It is used at 3 per cent by weight of dry soybeans (4) and coagulates at 78-85°C. A natural nigari solution with a relative density of 1.082 contains 16.5 per cent solids but it can be dried and used in the granular form.

b. **Sulphates.** Calcium sulphate or gypsum ($CaSO_4.2H_2O$) and magnesium sulphate or epsom salts ($MgSO_4.7H_2O$) are both used at 2.2 per cent of dry bean weight and at 70-75°C. Gypsum coagulates more slowly than the other coagulants to give the smoothest texture and maximum precipitation. The hydrated form is heated and then ground to fine particles before use.

c. **Acids.** Citrus juice, lactic acid or a 4 per cent solution of acetic acid (vinegar) are used. They lower the pH of the soya milk to 4.5 and the reaction is halted by neutralization with an alkali (eg sodium bicarbonate or sodium hydroxide). Lemon juice is used at 21 per cent of the weight of dry beans and vinegar at 16 per cent, at 80-90°C.

d. **Lactone.** Glucono delta-lactone (GDL) is mixed with soy milk at 0.6 per cent and heated to 85-90°C for 30-50 minutes producing gluconic acid which acts as a coagulant.

Modern methods of beancurd manufacture in China

A method has been developed and introduced which gives a high yield and no by-product. However the equipment is more sophisticated and its use is limited to those who can afford the higher investment and staff skills. The beans are soaked as previously described and frozen in liquid nitrogen at -70 to -100°C to make the cell membrane crisp and to disrupt the internal cell structure. When the beans regain a fresh appearance the freezing step is stopped. These very low freezing temperatures require cryogenic freezers with high capital expenditure and running costs so have limited use. During mashing the ratio of added water is 2 parts water to 1 part beans and they are mashed to a gel-like mixture a second time with the addition of warm water. A little cold water is added to the final thick soybean liquid so that the water is ten times the weight of the beans. Defoamers are sometimes used to eliminate bubbles, for example:

1) a mixture of equal parts calcium carbonate and edible oil,
2) glycerol monostearate used at 0.5-1 per cent of the weight of dry beans or,
3) a mixture of calcium hydroxide (78 per cent), calcium carbonate (12 per cent), magnesium carbonate (2 per cent), and acidified vegetable oil (8 per cent).

During the heating stage the soybean liquid is placed in a pan, flour is sprinkled on the surface and heated to boiling. The coagulant is added as usual but the mixture left for only 10 minutes to settle before draining and pressing. The most up-to-date plants use similar methods to the traditional process but more modern equipment. For example, the soybean mash is heated in automatic boilers under pressure and the entire process from grinding the beans to cooling the beancurd can take as little as 50 minutes. The traditional hand-operated screw presses or lever presses are replaced by hydraulic presses, the wooden boxes are up-dated by perforated aluminium containers and stone grinders, which are slow and labour intensive, give way to high speed grinders. Wood or charcoal fires are traditionally used for heating, whereas modern plants use propane burners or electric heaters. Changes in ingredients include the use of defatted soybean meal instead of whole soybeans to reduce the treatment time and a refined calcium sulphate is preferred. The defatted soybean meal can be made by extracting the liquid with hot hexane at 60°C for 30-40 minutes during which time there is no major loss of protein solubility, or enzymatic or trypsin inhibitor activity.

High yield beancurd

An increased yield and a low amount of by-product can be obtained by including a two-stage grinding step. For example 5kg of beans are soaked in 15kg of water with a further 15kg of water added during the first grinding stage and 7.5kg during the second stage. The machine should be washed with 5-6kg of water to avoid wastage. After boiling the soya milk for 2-3 minutes, no cold water is added, but when the temperature naturally falls to 80-90°C the coagulant is introduced. The usual pressing steps can then be carried out.

The list of products in China derived from the standard white soybean curd is almost endless, so an attempt is made here to mention the most common and interesting types. Their production involves such operations as smoking, drying, deep-frying and fermentation and they provide the customer with a wide range of tastes and textures.

Grinding the soaked beans to a mash.

Beancurd products

1: Deep-fried beancurd (You doufu).

During the manufacture of the standard white beancurd the liquid is only boiled to 95°C compared to the usual temperature of about 103°C for making the deep-fried varieties. The yield is lower hence 1kg of beans only produces 7kg of beancurd compared to the usual 9kg. The curd is cut into different shapes such as cubes (2cm high), strips (5x1x1cm^3) and triangles (1-1.5cm thick) (Figure 13) which are deep-fried in rapeseed or teaseed oil at 190°C until the surface is golden brown. The cubes are fried for only 35-45 seconds until the internal texture is light and contains many air pockets; the triangles are fried for 1 or 2 minutes and coated with seasonings. Unfortunately, since these varieties are sold by weight, poor quality products with a higher water content earn a higher profit.

Types of deep fried doufu.

Figure 13. Different shapes of deep-fried beancurd

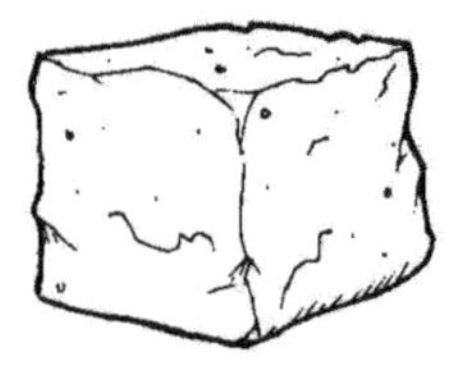

2: Deep-fried beancurd balls (Yuanzi).

This product is not widespread despite its simple manufacturing method and good taste. The recipe is:

standard white beancurd	500g
wheatflour	350g
starch	150g
salt	25g
onions (finely chopped)	to taste
chilli powder (dried)	to taste
ginger (chopped)	to taste

The ingredients are mixed well, formed into 2-3cm balls then deep-fried in a wok until golden in colour. They are sold by weight and reheated by the customer.

Deep fried cubes.

3: Dry/pressed beancurd (Doufu gan).

Many products are derived from this type of beancurd which has a much firmer texture than the standard product. The difference in its manufacture is first that 20-25 per cent more water is added to lower the soybean content and the coagulation speed, so that the water is easily removed on pressing. The second modification is that after the coagulant (calcium sulphate or magnesium chloride) is added, the liquid is mixed to form small particles which are left to settle. The upper layer of water is discarded leaving a slab about 7cm high which is pressed for about 1 hour until it is 1cm thick. This contains 22 per cent protein and 62 per cent moisture. This dry beancurd can be sold or treated further as below.

Sheets of pressed beancurd.

Shredded pressed beancurd.

3a: Beancurd cutlets (Nan hua gan).

A sharp knife is used to make shallow cuts diagonally across the top and bottom of each rectangular piece (20-30cm long, 6cm wide) as in Figure 14. The pieces are deep-fried quickly to harden the surface and then heated for 15 minutes in a solution of flavourings (eg chilli powder, salt, monosodium glutamate).

Figure 14. Cutting the curd

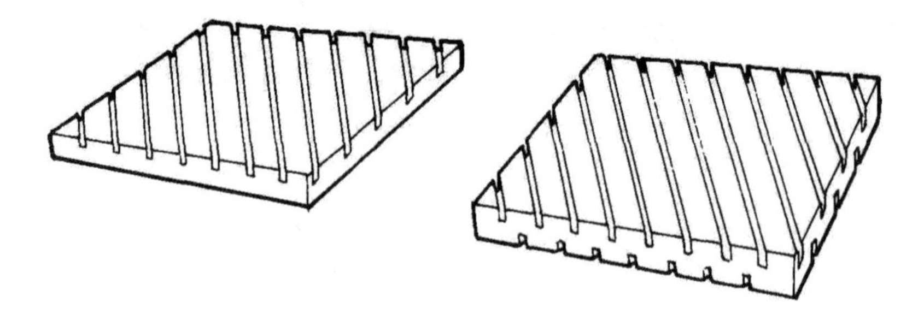

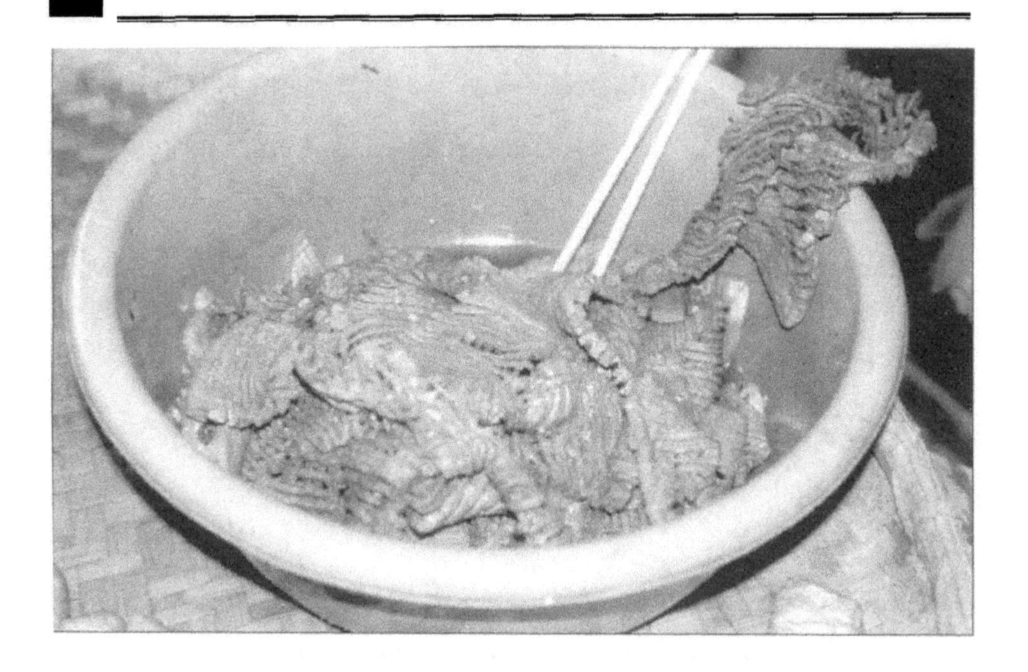

Beancurd cutlets.

Cutting into squares.

Cut squares before (front), and after smoking (back).

3b: Beancurd roll (Kunji).

This sausage shaped product (15cm long, 5cm diameter) is made by tightly rolling up a large rectangle of the dry beancurd, tying the ends, and boiling in a caramel solution to give a shiny brown colour.

Kunji.

3c: Smoked beancurd (Xun doufu).

The dry beancurd is cut into strips or 10cm square slabs which are laid on a shelf, below which is a 20cm layer of smoking rice bran (Figure 15). It is smoked slowly, turning it over when the surface is brown to give a distinctive flavour, colour and aroma. After smoking the texture is firm and the product can be stored for ten days without refrigeration due to the natural preservatives present in the smoke and the drying action (25 per cent weight loss by evaporation). Smoking can take several hours and may be cold smoking (below 30°C) or hot smoking (below 70°C).

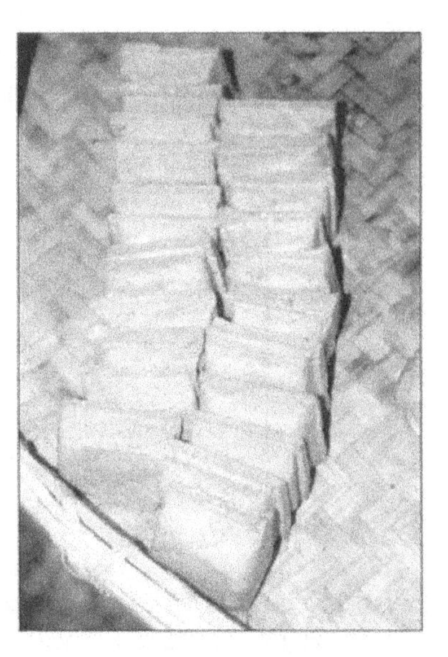

Smoked squares.

A highly flavoured product can be made by mixing 100kg dry beancurd, 0.05kg pepper, 0.05kg five-spice, 1kg sugar, 1kg alkali, 0.3kg monosodium glutamate, 2kg salt, 0.5kg chopped ginger and 4kg green Chinese onion plus a little water. This is then steamed for 10-12 minutes and mixed with 1kg vegetable oil. The product is spread on a bamboo mat, left in the sun to dry and then smoked to give a golden colour. Street stalls sometimes sell kebabs (Figure 16) made from smoked beancurd threaded onto a skewer, boiled in water to soften them, and flavoured by brushing on a spicy sauce containing onion, chilli powder, garlic, soy sauce, and sesame oil.

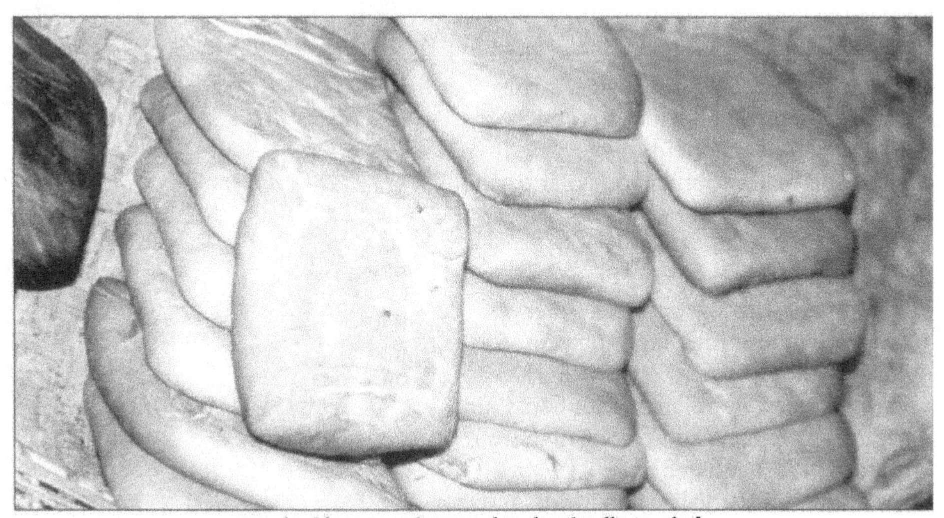

Smoked beancurd pressed individually in cloth.

Figure 15. Smoking equipment

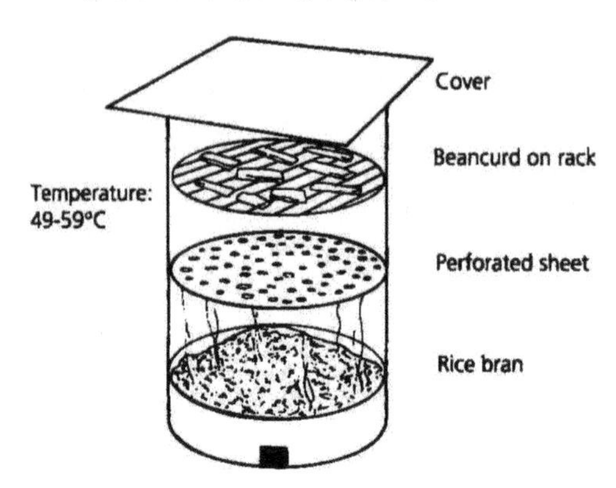

Figure 16. Beancurd kebabs

3d: Spiced beancurd (Xiang gan).

Soya beans are soaked in water containing salt, caramel and five-spice powder before being ground, cooked and filtered. A 25 per cent solution of coagulant is added to the stirred soybean milk and left for 15-20 minutes to coagulate. It is then pressed for 15-20 minutes in a cloth lined box before it is turned and cut into squares. Finally, a flavouring and colouring stage involves cooking for 20 minutes in a solution of 100g salt, 75g caramel powder, 50g five-spice and water to mix, and solar drying, three times.

A simpler method involves mixing the flavourings with the soya milk then coagulating and pressing. The pressed curd is cut into strips (3cm wide, 10cm long), cooked in boiling water containing caramel to harden and colour it, then dried. The flat squares have a brown surface and a hardish white inside and can be sliced thinly before stir frying.

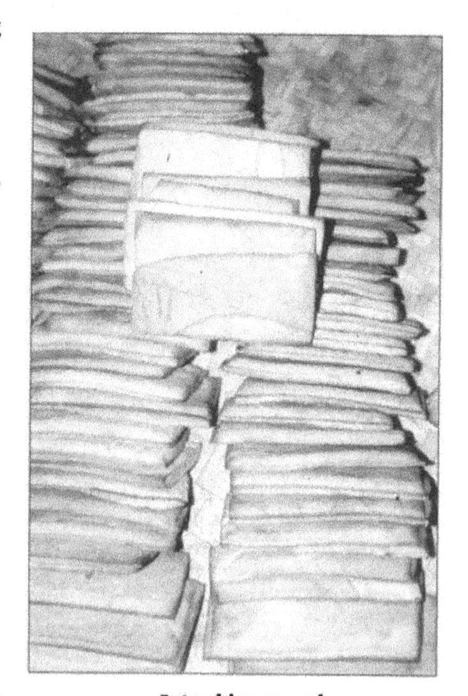

Spiced beancurd.

3e: Stewing spice beancurd (Luzhi doufu gan).

The stewing spice solution contains five-spice, soy sauce and vinegar cooked together. It can be used to make a spiced beancurd product by cutting the pressed curd into small pieces and preheating them by deep-frying in oil initially at 120°C and rising to more than 200°C with continued heating. When the temperature reaches 200°C the pieces are removed before being cooked in the stewing spice solution for approximately 2 hours. The ratio of fried beancurd to the solution is 1:2 or 2.5. After flavouring, it is cooled in a countercurrent tunnel with air speeds of less than 20-25 m/s and then packaged.

3f: Long life beancurd.

2-3mm thick slices are dried at less than 70°C to keep for many years under dry conditions. They can be eaten dried, deep-fried as a snack or rehydrated by soaking in hot water for 15-30 minutes.

4: Fermented beancurd (Furu).

This product has a soft, creamy consistency and a strong spicy, slightly alcoholic flavour somewhat resembling a soft, ripened cheese. Most families produce their own leading to many different tastes and qualities but it is also manufactured commercially and sold in most food shops. It is eaten as a relish, spread on steamed bread or with rice porridge. Beancurd with a moisture content of less than 70 per cent is cut into 8cm cubes which are traditionally spread on a bed of clean straw in a wooden box and with a 2cm gap between each piece. An inoculum of Mucor species is dissolved in cold boiled water and spread on the pieces which are then covered with plastic to retain the moisture. The box is left at 10-15°C (preferably below 13°C) for six days until a 5cm high layer of mould has grown on the surface. Using a chopstick this mould layer is spread over each cube and they are left in the sun to dry.

A liquor is made up of millet wine (alcohol 50 per cent v/v), chilli powder, monosodium glutamate, salt and ginger and is used to fill a jar of the fermented beancurd before it is sealed with wax or plastic and left for 10 days to mature. The proportion of flavourings varies with personal preference but usually 50g of millet wine is used per 100 pieces. Typically the bottled product will contain 32 pieces and have a net weight of 300g i.e. 200g of product in 100g of liquor. Commercially the beancurd is pasteurised, inoculated with Actinomycor elegans, Mucor species and Rhizopus species and left for 3-7 days at 20°C. The pieces are packed in brine

(12 per cent sodium chloride) and millet spirit and left to ferment before bottling in brine and undergoing a heat sterilisation process.

5: Beancurd bamboo (Fuzhu/dousun).

Dried yellow sheets of beancurd are rehydrated by the consumer in water at 40°C until soft, cut into small pieces and mixed with various flavourings (eg salt, soy sauce, ginger, sesame oil, vinegar and garlic). It is eaten as a cold dish, incorporated into soups or fried with vegetables and has an unusual, chicken-like texture. It is also sometimes used to wrap other sweet and savoury foods. The beancurd bamboo is made by heating soya milk in a flat, shallow, open pan until at about 60°C a skin forms on the surface. It is removed by lifting with a rod and hung on a rack to dry. As new films form, they are successively removed until no more form; the films have lower protein and fat contents than the dried product and increased carbohydrate and ash contents. The dried product (2 per cent moisture) is brittle and due to its high lipid content (24 per cent) has a fairly short shelf life.

Hydrated spiced bamboo.

Beancurd bamboo.

A Chinese patent (5) describes a modern method of production. To summarise: 100kg of soybeans are dehulled, ground and added to 55-60kg water containing 0.6-1.5kg sodium chloride and 0.3-0.6kg sodium carbonate. They are blended for 30 seconds in a high speed mixer and the mash is fed into a specially developed twin-screw extruder. After leaving the extruder the final moisture content is 10 per cent and the yield is approximately 80-90 per cent. This method greatly speeds up the process so that the time from receipt of the beans to leaving the extruder is only 80-90 minutes. The specifications for the extruder are an inlet temperature of 70-95°C, a central barrel temperature of 150-190°C and an outlet temperature of 140-160°C (pressure 20-30 kg/cm^2). A further modification to this process is to feed a mixture of defatted soybean powder, water (45-52 per cent) and spices into the extruder at 80-90°C (central barrel temperature 140-180°C). Different shaped products arise from various dies giving tubes, strips or threads 0.5-1mm wide, which can be used fresh. The composition of the final product is: fat 5-25 per cent, protein >45 per cent, ash <6 per cent, fibre <2 per cent. The expansion in product volume is three times.

6: Frozen beancurd (Dong doufu).

In the winter beancurd can be left outside for several weeks to freeze and develop a spongy, chewy texture. It is thawed in warm water, pressed and dried for storage. The dried curd can be rehydrated with cold water. This traditional method has been commercialised by quick freezing pieces of beancurd at -10°C then holding them at -3°C for 20 days. The frozen pieces are thawed with a water spray and pressed to facilitate the drying step that follows. The pressed curd is sprayed with a solution of baking powder, pressed again and dried at 100°C for 15-20 hours. The total time required is 23 days. The composition of the product is: protein 63.4 per cent, fat 26.4 per cent, carbohydrate 7.2 per cent, water 0.4 per cent, other matter 2.6 per cent.

7: Flavoured beancurd.

It is reported (6) that a number of new flavoured beancurd products have been developed in Japan, by adding different ingredients to the soya milk before it is coagulated. The variations include the following examples:

a. **Fruit flavoured curd.** By adding acidic fruit juices the process of coagulation is facilitated and they also provide fruit flavours. The pH of the fruit juice should first be adjusted to 2.3-2.9 and 14-35ml are added to 11 litres soybean milk.

b. **Spiced beancurd.** Hot chilli powder is soaked in hot water and mixed with the soybean milk before the calcium sulphate is added.

c. **Milk beancurd.** Acetic acid is added to cow's milk to separate the milk proteins which are then added to soybean milk. The proportion of acetic acid to milk is 0.02:100 and the ratio of milk proteins to soybean milk is 3:7.

d. **Pork beancurd.** Dried ground pork with an emulsifying agent is added to the soybean milk before the curd is manufactured.

e. **Coffee beancurd.** 0.5-10 per cent soluble coffee and sugar (optional) are added to the soybean milk at 40°C, which is coagulated with GDL. The product is packed and heat-processed in 300ml pots by heating in water at 90°C for 50 minutes.

8: Specialities.

Every region has its own locally developed products which have become regional specialities. There are too many to list, and the examples given represent a few of the many different products available.

a. **Jinxi ginger beancurd (7).** Pieces of the dry/pressed beancurd are boiled for 5 minutes in a solution of finely chopped root ginger; the ratio of the beancurd to ginger is 5:2. The beancurd is not pressed in boxes, as in the usual method, but is wrapped in cloth and pressed on a wooden table (Figure 17). Finally the flavoured curd is dried over charcoal.

b. **Shaoyang pork roll.** The 'pork roll' has a solid consistency and can be steamed to soften then sliced and eaten directly or stir-fried. It has a distinctive tangy flavour and the slices are bright red in colour with a black rind. It is made by mixing 80 per cent white beancurd, 15 per cent minced pork and 5 per cent pork blood, chilli powder, grated orange peel and seasonings. The mixture is shaped by hand into balls (400g each) which are smoked very slowly until the surface becomes black and each piece is about 7cm long and weighs 50g.

Figure 17. Pressing on a wooden table

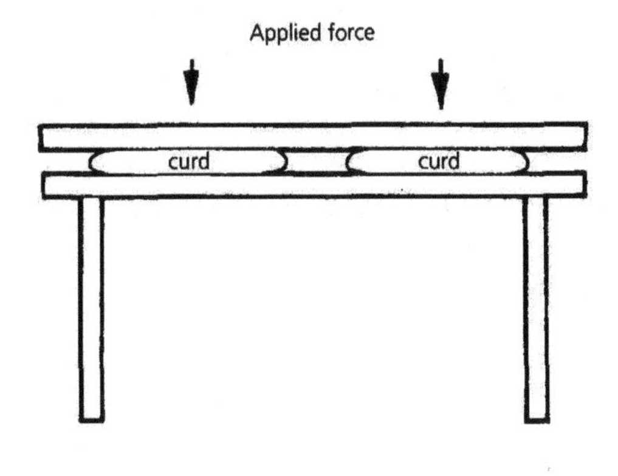

c. **Bean paste (douhu).** A paste can be made which is eaten directly or used to make a beverage by mixing 20g unsweetened paste, 25g sugar and 175ml boiling water. The paste is made from 100kg soybeans soaked for 45 minutes in water at 80°C. It is heated with steam to raise the temperature to 100°C within 20 minutes, and boiled for 10 minutes. After draining, 32kg of product is put into a frying pan with 27 litre of oil at 180°C and heated for 5 minutes after the product temperature reaches 160°C. 1kg of fried product is mixed with 100g palm oil and ground to a paste which can be sweetened with 300g sugar. The composition per 100g of paste is as follows:

	Before Grinding	After Grinding
Water (g)	4.0	1.2
Crude Protein (g)	36.8	28.4
Fat (g)	38.9	51.7
Carbohydrate (g)	17.8	16.4
Crude Fibre (g)	2.9	-
Ash (g)	3.1	2.4
Calcium (mg)	138	157
Phosphorus (mg)	255	357
Iron (mg)	5	-

9: Beancurd silk.

New beancurd products, such as beancurd silk, are continually being researched and developed in China (8). The product is in the form of long strands, hence the name silk, and can be consumed in different ways. For example, deep-fry in soybean oil at 180°C until golden brown, soak the strands in boiling brine (the ratio of water:sodium chloride is 7:3) for 30 minutes, spread on an oiled tray to dry and give a golden colour, eat as a snack. To improve the flavour these golden strands can be heated in a pre-boiled stock (7.5kg water, 2kg pork bone, 1kg chicken, spices) for 10-15 minutes before being packaged in metal foil and pasteurised at 116°C for 20 minutes.

To make the strands, soybeans are soaked in 3 times their weight of water at 20-30°C for 5 hours. The milk remaining after grinding and filtering is heated to 97°C, then cooled to 85-90°C before adding the coagulant (80g $CaSO_4$ in 1.5-2kg soya milk). After 25 minutes the tender curd is ready to be pressed in a wooden box lined with cloth, and layered between the folds of a 10 metre length of cotton (Figure 18) until each layer is 0.1-0.15cm thick. The silk-like strands are produced by cutting into strips 0.1-0.2cm wide and 4cm long.

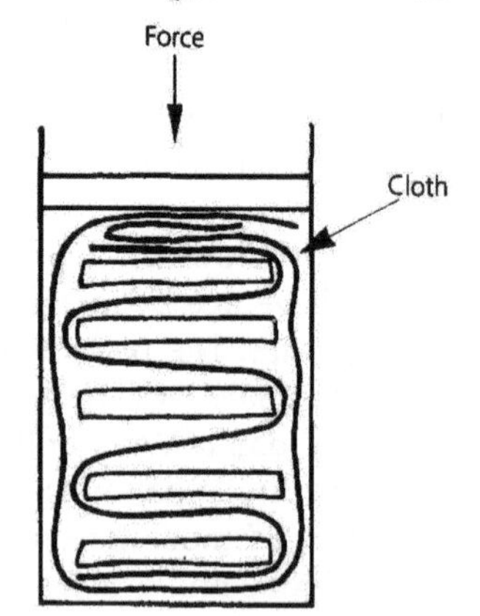

Figure 18. Layering the curd

10: Heat processed beancurd.

Soya beans can be used to make a curd which is packed in 300g plastic containers, sealed and heat processed in water at 85°C for one hour (9). The procedure involves crude soya milk (ie ground, soaked, dehulled beans) which undergo steam treatment and centrifugation before being homogenised at a pressure of 150kg/cm^2. The following phosphate compounds are added to the milk: 0.084 per cent anhydrous sodium pyrophosphate ($Na_4P_2O_7$), 0.058 per cent sodium tripolyphosphate ($Na_5P_3O_{10}$), 0.058 per cent sodium hexametaphosphate ($NaPO_3$), which is then cooled to 10°C and coagulated with 0.5 per cent magnesium chloride ($MgCl_2$).

11: Instant doufu mix.

A dried powder that can be used to make beancurd simply by adding water is a relatively new product. The mix is made by taking 10kg defatted soybean powder, mixing with 100 litres of water at 50°C and centrifuging. The by-product which settles out is discarded. The upper layer is adjusted to pH 4.5 with hydrochloric acid giving approximately 3.2kg of sediment which is diluted with water so that the solids content is 7 per cent. 1N sodium hydroxide is added to adjust it to pH 6.8. The liquid is heated with steam at 120°C and 2.4kg of palm oil is added. The next stage is to homogenise and spray dry the liquid. One litre of water is mixed with 200g of the dried product for 15 minutes at 1500 rpm at room temperature. Finally, the product is frozen at -40°C and vacuum dried.

Use of by-products

The by-products from standard beancurd manufacture (Figure 2) are: soybean hulls; pulp remaining after the mash is filtered; and whey after the curds are drained. Attempts are made to utilize the hulls and pulp to avoid wastage and increase profits.

Hulls are used to feed livestock or as a fuel. When dried and ground the powder contains 10 per cent protein, 38 per cent crude fibre and 38 per cent other carbohydrates which can be added as a nutritional supplement to baked products at a rate of 5-50 per cent.

By-product after filtration.

The composition of the pulp is shown in Figure 19. The moisture content is 87 per cent and after partial drying can be lowered to 75 per cent. Based on research carried out in Japan, pulp can also be dried and ground for use as an ingredient in a number of products. Figure 20 shows the composition of the protein rich product compared to wheat flour.

Figure 19. Composition of the pulp

Moisture Content (%)	Protein (%)	Fat (%)	Carbohydrate (%)	Crude Fibre (%)	Ca (mg)	P (mg)	Fe (mg)	Vitamin B1 (mg)	Vitamin B2 (mg)
87.0	2.6	0.3	7.6	1.8	16	44	4	0.03	0.02
75.0	5.1	0.6	14.9	3.5	13.3	86.1	7.8	0.15	0.03

Figure 20. Composition (per 100g) of dried ground pulp compared with wheat flour

	Protein (g)	Fat (g)	Carbohydrate (g)	Crude fibre (g)	Ca (mg)	P (mg)	Fe (mg)	Vitamin B1	Vitamin B2	Nicotinic Acid
Wheat Flour	10	0.7	77.7	0.1	22	112	-	-	0.07	1.5
Ground Pulp	20	2.3	58.5	15	123	338	30.8	0.6	0.15	0.8

Examples of the way by-products are utilized are as follows:

a. In extruded potato snacks, 20 per cent of the potato starch is substituted by pulp which has been dried and ground to particles of 0.15mm-0.2mm. After frying the product is said to be more acceptable than one using 100 per cent potato starch.

b. Since sponge cakes made with corn starch are usually too soft, adding ground pulp to replace 10-50 per cent of the corn starch improves the texture.

c. A fried cake can be made by mixing and then grinding the following ingredients: 12kg untreated pulp, 4kg starch, 1kg wheat flour, 1.8kg sugar, 100g salt and 50g sesame seeds. It is not necessary to add water since the fresh by-product contains 80 per cent water. If, however, the consistency is too dry, 0.5-1kg of soybean liquid (ie boiled, ground soybeans) can be added. The mixture is rolled and cut into lengths before deep-frying.

d. Wheat which is added during the initial stages of soy sauce production can be replaced by the dried pulp. If 10kg of the pulp is mixed with 16kg bran, water must be added to give a final moisture content of 40 per cent. The next stage is to heat to 110°C for 30 minutes, cool to 40-45°C and shape into cylinders approximately 40-50mm long and 20-30mm in diameter. They are inoculated with Aspergillus oryzae, fermented for 40 hours and cut into pieces. These are mixed well with 40kg water and 6kg salt and can be stored for use in soy sauce manufacture.

e. The dietary fibre can be extracted by adding acid to the by-product then by heating under high pressure, washing and drying. The washing stage dissolves the polysaccharides. The resulting white powder is soybean dietary fibre, free from 'beany' odour, with an expansion rate of 3.5 times in water. The composition is 30 per cent insoluble dietary fibre, 18 per cent protein and 8 per cent fat.

Cooking with beancurd

Beancurd in its many forms is a popular, nutritious and fairly inexpensive raw material available throughout China. Every region has its own specialities from spicy mapo doufu made in Sichuan Province to oyster sauce beancurd in Guangdong Province. It was traditionally the basis of a family dish but now it is often an important ingredient of the wonderful delicacies served at banquets. Before cooking the white beancurd may be treated to remove the beany odour by rinsing small cubes in water then parboiling for 1 to 5 minutes (11). If it is boiled for 30 minutes it becomes soft and has a very open texture due to the formation of many small holes. Alternatively it can be steamed in one of two ways; the first being in a pot so that the steam saturates the ingredients, and the second in a double boiler. As mentioned earlier, freezing removes some water and gives a firm texture.

Recipe details are out of the scope of this booklet and are too numerous to cover. Recipe books are available in English especially books designed for vegetarian cuisine (11).

References

1. Wang Wen Qiao and Gang Wen Bin, *Chinese vegetarian cuisine,* New World Press. (1990).

2. H E Snyder and T W Kwon, *Soybean utilization,* Van Nostrand Reinhold Co. (1987).

3. Thio Goan Lao, *Small-scale and home processing of soya beans with applications and recipes,* Communications 64(a) Royal Tropical Institute, Netherlands.

4. W Shurtleff, *Tofu and soy milk products, the book of tofu,* Vol.VII, Soyfoods Centre (1979).

5. *Doufu pi and fuzhu production,* Patent CN86100825A. (In Chinese).

6. *New japanese tofu products,* Vol. 4, Translated into Chinese by Xia Heng Nian; J of Food Industry (1991). (In Chinese).

7. Liu Bao Jia, *Comprehensive food processing technology and recipes.,* Vol. 1, Science and Technology Reference Publications. (In Chinese).

8. *Cooked doufu silk production,* Patent 88101315.3 (22.12.88). (In Chinese).

9. *A new doufu product manufacture,* Patent CN85105100A. (In Chinese).

10. Fang Tao, *By-product of soybean processing,* Food Information, September (1990). (In Chinese).

11. Zhang De Sheng, *200 recipes for the doufu devotee,* Women of China, Beijing Press. (1986).

Acknowledgements

The work reported here was undertaken while the author was working at Human Agricultural University, PRC. The encouragement and co-operation of the staff of its Faculty of Food Science is warmly acknowledged, particularly Tang Shuze and Lin Qin Lu. Considerable thanks are also due to United Biscuits (China) for the provision of a generous grant towards the production costs of this article.

Thanks are also due to Intermediate Technology Development Group: to Dr Peter Fellows for editorial work, to the Publishing Unit for preparation and layout, to Matthew Whitton for drawings and to the Overseas Development Administration of the British Government for financial support.

Intermediate Technology
Myson House
Railway Terrace
Rugby
CV21 3HT

Tel: 0788 560631
Fax: 0788 540270

Intermediate Technology Development Group Ltd., Patron: HRH The Prince of Wales, KG, KT, GCB.
Company Reg. No. 871954, England. Reg. Charity No. 247257.
Printed by Neil Terry Printing on Brandblade ★★★★ and Lumiart ★★★★★

JN/2216/94/02